Limiting Armed Drone Proliferation

COUNCIL on FOREIGN RELATIONS

Center for Preventive Action

Council Special Report No. 69
June 2014

Micah Zenko and Sarah Kreps

Limiting Armed Drone Proliferation

The Council on Foreign Relations (CFR) is an independent, nonpartisan membership organization, think tank, and publisher dedicated to being a resource for its members, government officials, business executives, journalists, educators and students, civic and religious leaders, and other interested citizens in order to help them better understand the world and the foreign policy choices facing the United States and other countries. Founded in 1921, CFR carries out its mission by maintaining a diverse membership, with special programs to promote interest and develop expertise in the next generation of foreign policy leaders; convening meetings at its headquarters in New York and in Washington, DC, and other cities where senior government officials, members of Congress, global leaders, and prominent thinkers come together with Council members to discuss and debate major international issues; supporting a Studies Program that fosters independent research, enabling CFR scholars to produce articles, reports, and books and hold roundtables that analyze foreign policy issues and make concrete policy recommendations; publishing Foreign Affairs, the preeminent journal on international affairs and U.S. foreign policy; sponsoring Independent Task Forces that produce reports with both findings and policy prescriptions on the most important foreign policy topics; and providing up-to-date information and analysis about world events and American foreign policy on its website, CFR.org.

The Council on Foreign Relations takes no institutional positions on policy issues and has no affiliation with the U.S. government. All views expressed in its publications and on its website are the sole responsibility of the author or authors.

Council Special Reports (CSRs) are concise policy briefs, produced to provide a rapid response to a developing crisis or contribute to the public's understanding of current policy dilemmas. CSRs are written by individual authors—who may be CFR fellows or acknowledged experts from outside the institution—in consultation with an advisory committee, and are intended to take sixty days from inception to publication. The committee serves as a sounding board and provides feedback on a draft report. It usually meets twice—once before a draft is written and once again when there is a draft for review; however, advisory committee members, unlike Task Force members, are not asked to sign off on the report or to otherwise endorse it. Once published, CSRs are posted on www.cfr.org.

For further information about CFR or this Special Report, please write to the Council on Foreign Relations, 58 East 68th Street, New York, NY 10065, or call the Communications office at 212.434.9888. Visit our website, CFR.org.

Printed in the United States of America.

To submit a letter in response to a Council Special Report for publication on our website, CFR.org, you may send an email to CSReditor@cfr.org. Alternatively, letters may be mailed to us at: Publications Department, Council on Foreign Relations, 58 East 68th Street, New York, NY 10065. Letters should include the writer's name, postal address, and daytime phone number. Letters may be edited for length and clarity, and may be published online. Please do not send attachments. All letters become the property of the Council on Foreign Relations and will not be returned. We regret that, owing to the volume of correspondence, we cannot respond to every letter.

Contents

Foreword

The United States is the world's most prolific user of armed drones; indeed, Israel and the United Kingdom are the only other countries known to have utilized such weapons in combat. Yet it would be unwise for Washington to ignore the risk of armed drone proliferation. Though surveillance drones are in wide use around the world, countries seeking armed drones are often in areas of tenuous security, where a new weapons system can be inherently destabilizing. China and Iran are already thought to have functioning armed drones. India, Pakistan, Turkey, and others have expressed an interest in acquiring them. Even Hezbollah, the Iranian-supported terrorist group, attempted to use them during its 2006 war with Israel. And all of this activity is taking place in the absence of commonly adhered-to regulations and norms on the sale and use of these weapons.

In this Council Special Report, Senior Fellow Micah Zenko and Stanton Nuclear Security Fellow Sarah Kreps argue that it is essential to begin working now to expand and establish such rules and norms, while the number of states with armed drones remains relatively small. Doing so has some potential to reduce the odds that armed drones get into the wrong hands anytime soon. Even so, the unique ability of drones to hover for long periods over a target and react quickly to strike opportunities, all with no risk to a pilot, means, the authors believe, that they will be deployed more frequently than other armed assets. This has the potential to raise the number of armed interactions among states and increase—perhaps dangerously—the costs of misinterpretation and miscalculation on the part of governments.

To minimize the scale of armed drone proliferation and to mitigate some of its risks, the authors call on the United States to take the lead in efforts to expand the reach and comprehensiveness of existing drone proliferation regulations and help establish, through its own behavior, norms governing their use. Should it prove impossible to reach agreement on a new proliferation treaty specifically addressing drones, which the authors recognize may be the case, they recommend that the United States limit its advanced armed drone sales only to those countries that commit to basic protections of human rights, peaceful resolution of disputes, and the nonproliferation of weapons of mass destruction. They also outline specific, limited modifications to existing regulations that could be helpful in the absence of more comprehensive controls.

Zenko and Kreps maintain that the United States, as the main exemplar of drone use, has a unique responsibility and opportunity to demonstrate norms for drone use. They encourage Washington to be more transparent about its targeting decision-making and more specific as to the domestic and international legal constraints that govern its drone operations. By doing so, they predict, the United States will create standards of behavior that other countries will be more likely to follow.

Limiting Armed Drone Proliferation raises an important set of issues. It describes the growing scale of armed drone proliferation and the risks it may raise, assesses ongoing debates regarding drone governance, and makes thought-provoking recommendations for future action. The result is a valuable report that is well worth reading and considering.

Richard N. Haass
President
Council on Foreign Relations
June 2014

Acknowledgments

We would like to express our gratitude to the many people who made this report possible. To begin, we would like to thank CFR President Richard N. Haass and Director of Studies James M. Lindsay for their support of this project and insightful feedback throughout the drafting process.

The report's advisory committee was an invaluable resource. In particular, we are especially grateful to members of the committee who went above and beyond the call of duty—namely, Sarah Knuckey, Michael Horowitz, and Steven J. Zaloga. We owe a huge debt of gratitude to Andrew Shapiro, who chaired the advisory committee. Additionally, Brian C. Babcock-Lumish, Daniel Byman, Robert M. Chesney, Stephen J. Del Rosso Jr., Ian Easton, James T. Golby, Dennis M. Gormley, Robert O. Keohane, Matthew Leatherman, Joanne Mariner, Daniel S. Markey, David A. Merkel, Scott Osterling, Ely Ratner, Paul B. Stares, Ward Thomas, Erik Voeten, and Graham Warwick provided input and critiques, all of which greatly improved the substance of the report.

The report also benefited from countless interviews conducted with current and former civilian and military officials from the U.S. government, as well as insights from the growing number of researchers, journalists, and activists concerned about proliferation of armed drones.

We are grateful for the valuable assistance of Patricia Dorff, Eli Dvorkin, and Ashley Bregman in CFR's Publications Department, who provided unmatched editing support, and to Cathryn Adams and Melinda Wuellner in Global Communications and Media Relations for their outstanding marketing efforts. We also appreciate the contributions of the David Rockefeller Studies Program staff, including Amy Baker, in shepherding the report.

Tremendous thanks are owed to colleague and Director of the Center for Preventive Action (CPA) Paul B. Stares, who provided needed guidance and insights. We are also grateful for the logistical and research support of CPA interns Priscilla Kim and Julie Anderson, and are especially appreciative of the essential role filled by Research Associate Amelia M. Wolf, without whom the report would not have been completed.

This publication was made possible by a grant from Carnegie Corporation of New York. The statements made and views expressed herein are solely our own.

Micah Zenko
Sarah Kreps

Council Special Report

Introduction

The use of unmanned aerial systems—commonly referred to as drones—over the past decade has revolutionized how the United States uses military force. As the technology has evolved from surveillance aircraft to an armed platform, drones have been used for a wide range of military missions: the United States has successfully and legitimately used armed drones to conduct hundreds of counterterrorism operations in battlefield zones, including Afghanistan, Iraq, and Libya. It has also used armed drones in non-battlefield settings, specifically in Pakistan, Yemen, Somalia, and the Philippines. Collectively, these strikes have eliminated a number of suspected terrorists and militants from Asia to Africa at no cost in terms of U.S. casualties, an advantage of drones over manned platforms that has made them attractive to many other states. However, non-battlefield strikes have drawn criticism, particularly those conducted under the assertion that they are acts of self-defense.

Though the United States remains the lead actor in terms of possessing and using armed drones, the rest of the world is quickly catching up. Russia, China, Iran, South Korea, and Taiwan, for example, have begun to develop increasingly sophisticated indigenous drone capabilities. Other countries, including Pakistan, Turkey, Saudi Arabia, and the United Arab Emirates (UAE), have publicized their intent to purchase them.

The direct consequences of armed drone proliferation on U.S. national security are several years out, but the policymaking decisions that will shape those consequences confront the Obama administration today. How the United States uses armed drones and for what purposes will contribute to the norms that will influence how states use them in the future. Under the leadership of the United States, norms regarding the proliferation and use of weapons—from nuclear and biological weapons to blinding lasers and antipersonnel landmines—have been overwhelmingly adopted and followed. Similar efforts should be made for the proliferation and use of armed drones, even if not all states abide by these norms. U.S. export policy will determine, to a certain extent, which states acquire what types of armed drones, and will set expectations about appropriate exports by other armed drone producers. If the United States reinforces multilateral institutions designed to limit armed drone proliferation, it will have the ability to shape the constraints that other states will face when acquiring drones.

Persistent media attention tends not to differentiate between armed and commercial drones, but rather homogenizes all types, despite the fact that armed drones will be more destabilizing. Though the armed drones acquired by states in the near term likely will not have capabilities equal to those of the United States, their effects will still be destabilizing. States that acquire armed drones will likely use them as probes and for limited attacks in international waters and across borders, against domestic threats, and, potentially, for even more lethal missions, including delivering weapons of mass destruction (WMDs). Although other vehicles, such as trucks and manned civilian aircraft, can also be used to deliver WMDs, the ability of drones to hover and wait for the opportune moment in which they can produce maximum effect confers uniquely lethal capabilities. If the United States delays and forgoes the opportunity to establish rules of the road for the use of armed drones and to constrain their proliferation, there will be grave consequences for U.S. interests, in terms of the prevention of armed conflict, promotion of human rights, strengthening of international norms and legal frameworks, and the future of warfare.

Subsequently, the United States should pursue a strategy that limits the proliferation of armed drones and promotes their use in a manner consistent with international law and norms, and that does not threaten U.S. interests or allies. The strategy should consider foreseeable destabilizing or deadly missions over the next decade and beyond, but remain flexible enough should unprecedented uses and missions emerge. Such a strategy will be difficult to execute and require sustained high-level attention from the Obama administration and its successors. Given that many states want to acquire armed drones, and drone producers outside the drone transfer regime, such as China and Israel, are already exporting drones, the need for implementation is all the more pressing.[1] Such a strategy would serve U.S. national interests in the following ways:

- Minimize the proliferation of the most capable and lethal drones to countries that are conflict prone.
- Reduce the potential for militarized disputes between states that could lead to an escalation of armed conflict in unstable regions.
- Decrease the likelihood that states and nonstate actors will use armed drones against the United States and its allies.
- Establish a more widely accepted legal and operational basis for conducting drone strikes to ensure that countries do not use armed drones in ways that threaten the United States or its allies.
- Increase the likelihood that internationally accepted norms and rules for armed drone exports and use will be adopted by emerging drone powers.

Drone Proliferation Trends

Analyzing which countries are pursuing armed drones is difficult, as their development is shrouded in secrecy and misinformation. Some countries, including the United States, hide certain programs to protect sensitive information and capabilities, while others, such as Iran, boast of armed drones to garner national prestige, despite the fact that they have not been demonstrably tested or used. In addition, government announcements of deadlines for internal drone development often go unmet, and publicly proclaimed export orders are never fulfilled.

The United States attracts the greatest attention because it is, by far, the most prolific user of armed drones. The British military, however, was responsible for 299 drone strikes in Afghanistan through July 2013.[2] Israel has used drones both in and outside armed conflicts as well. During the 2006 Lebanon War, both Israel and Hezbollah were reported to have used drones. Israel conducted a strike against suspected Hezbollah militants in southern Lebanon on July 31, 2006, while Hezbollah was reported to have deployed four Iranian-made drones toward Israel, though none succeeded in its mission.[3] Additionally, Israeli drones were used to conduct an estimated forty-two strike missions in the 2009 Gaza conflict, according to a joint investigation by Israeli and Palestinian human rights organizations, and to conduct cross-border attacks targeting suspected terrorists—such as the August 2013 strike carried out in the Sinai Peninsula with the consent of the Egyptian government.[4]

Many countries are positioning themselves to have the ability to emulate these capabilities. Other than the United States, Britain, and Israel, only China and Iran are believed to have operationally deployed armed drones, but they have not conducted strikes; both countries have showcased their capabilities and claimed that they are prepared to use them during a crisis. Iran claims to have developed a multi-role unmanned platform with a range of up to two thousand kilometers, which could potentially overfly much of the Middle East.[5] China has displayed what it claims are armed drones to the media for half a decade, and its spending on drones is surging at such high rates that it will equal that of the United States by 2020.[6]

According to industry estimates, international interest in armed drones has grown in the wake of Iraq and Afghanistan. The drone market is expected to grow from $5.2 billion in 2013 to $8.35 billion by 2018.[7] While drones are still a relatively small portion of the overall defense market, the segment with the "biggest potential" is the demand for medium-altitude long-endurance (MALE) drones, such as the Predator and Reaper.

Consistent with these growth trends, several other countries have announced their own armed drone programs. India reports that it will soon equip its drones with precision-guided munitions and hopes to mass-produce combat drones to conduct targeted strikes in cross-border attacks on suspected terrorists.[8] Rebuffed by requests to procure U.S. armed drones, Pakistan said it will develop them indigenously or with China's help to target the Taliban in its tribal areas.[9] The South African firm Denel aspires to sell armed variants of its Seeker 400, but only to governments that use them accountably and responsibly against "opportunistic" targets, which a company spokesperson characterized by saying: "That target could be a pirate or could be a terrorist."[10] According to Oxford Research Group, Turkey has about twenty-four types of drones in use or development, four of which have been identified as combat drones.[11] Switzerland and several European Union (EU) member states—including France, Italy, Spain, Greece, and Sweden—have collaborated on the Neuron, a stealth armed drone that made its first demonstration flight in December 2012. It is merely a technology demonstrator, however, and the stealth and weaponization components are still in development.[12] Additionally, the EU's goal to collectively develop a MALE armed drone remains elusive despite years of discussions.[13]

The final category comprises the majority of all aspiring drone countries that seek only unarmed surveillance drones while retaining the option to pursue more advanced military purposes later. For example, in 2012, the Australian military reported that armed drones remain "an option for a future force."[14] A number of Asian countries, including Japan and Singapore, are pursuing Global Hawk surveillance drones to monitor their borders, but have not yet indicated an interest in armed drones.

How Drones Are Different and Can Destabilize

Drones should be treated as a distinct class of weapons. They have unique properties that lead them to be used, and defended against, in ways that are destabilizing. In November 2013, Canada's chief of defense staff, General Thomas Lawson, made a claim that is commonplace among military officials: "If a kinetic round is propelled toward a confirmed enemy for strategic purposes by a rifle, by an artillery piece, by an aircraft manned, or by an aircraft unmanned, any of those that end up with a desired effect is a supportable point of view."[15] Similarly, in May 2012, then chief of staff of the U.S. Air Force, General Norton Schwartz declared, "If it is a legitimate target, then I would argue that the manner in which you engage that target, whether it be close combat or remotely, is not a terribly relevant question."[16] This premise is technically true and consistent with military officials' efforts to demystify drones, reduce public opposition to them, and integrate them into their armed services. However, these assumptions overlook the unique advantages of armed drones, which raise the prospect for moral hazard, where governments are more willing to use them over other weapons platforms because the associated costs and risks are assumed to be comparably lower. Indeed, the Obama administration commissioned a separate review of U.S. drone export policies, precisely because the vast majority of U.S. officials believe that it is a distinct weapons system.[17]

The attractiveness of armed drones stems from three inherent advantages. First, the typical MALE drone can persist over a target for up to fourteen hours without being refueled, which provides a continuous monitoring of the situation below. In contrast, manned aircraft can neither loiter nor fly repeatedly over an area of interest for more than four to six hours due to fuel or pilot limitations. Second, with a missile attached to the surveillance platform, the responsiveness of armed drones when time-sensitive targets appear in the operator's view is unmatched. Moreover, drone-fired missiles can be diverted by the weapons system operator at the last moment if noncombatants enter the likely blast radius.[18] Third, and most important, unmanned systems do not place human pilots or ground forces at risk of being killed or captured in hostile environments. Such advantages have made drones the "weapon of choice" of the United States for killing suspected terrorists.[19]

The inherent advantages of drones will not alone make traditional interstate warfare more likely—such conflicts are relatively rare anyway, with only one active interstate conflict in both 2012 and 2013.[20] Nor will the probable type, quantity, range, and lethality of armed drones that states possess in coming decades make a government more likely to attempt to defeat an opposing army, capture or control foreign territory, or remove a foreign leader from power. However, misperceptions over the use of armed drones increase the likelihood of militarized disputes with U.S. allies, as well as U.S. military forces, which could lead to an escalating crisis and deeper U.S. involvement. Though surveillance drones can be used to provide greater stability between countries by monitoring ceasefires or disputed borders, armed drones will have destabilizing consequences. Arming a drone, whether by design or by simply putting a crude payload on an unarmed drone, makes it a weapon, and thereby a direct national security threat for any state whose border it breaches.

INCREASED FREQUENCY OF INTERSTATE AND INTRASTATE FORCE

For the United States, drones have significantly reduced the political, diplomatic, and military risks and costs associated with the use of military force, which has led to a vast expansion of lethal operations that would not have been attempted with other weapons platforms. Aside from airstrikes in traditional conflicts such as Libya, Iraq, and Afghanistan—where one-quarter of all International Security Assistance Force (ISAF) airstrikes in 2012 were conducted by drones—the United States has conducted hundreds in non-battlefield settings: Pakistan (approximately 369), Yemen (approximately 87), Somalia (an estimated 16), and the Philippines (at least 1, in 2006).[21] Of the estimated 473 non-battlefield targeted killings undertaken by the United States since November 2002, approximately 98 percent were carried out by drones. Moreover, despite maintaining a "strong preference" for capturing over killing suspected terrorists since September 2011, there have been only 3 known capture attempts, compared with 194 drone strikes that have killed an estimated 1,014 people, 86 of whom were civilians.[22]

Senior U.S. civilian and military officials, whose careers span the pre– and post–armed drone era, overwhelmingly agree that the threshold for the authorization of force by civilian officials has been significantly reduced. Former secretary of defense Robert Gates asserted in October 2013, for example, that armed drones allow decision-makers to see war as a "bloodless, painless, and odorless" affair, with technology detaching leaders from the "inevitably tragic, inefficient, and uncertain" consequences of war.[23] President Barack Obama admitted in May 2013 that the United States has come to see armed drones "as a cure-all for terrorism," because they are low risk and instrumental in "shielding the government" from criticisms "that a troop deployment invites."[24] Such admissions from leaders of a democratic country with a system of checks and balances point to the temptations that leaders with fewer institutional checks will face.

President Obama and his senior aides have stated that the United States is setting precedents with drones that other states may emulate.[25] If U.S. experience and Obama's cautionary words are any guide, states that acquire armed drones will be more willing to threaten or use force in ways they might not otherwise, within both interstate and intrastate contexts.

States might undertake cross-border, interstate actions less discriminately, especially in areas prone to tension. As is apparent in the East and South China Seas, nationalist sentiments and the discovery of untapped, valuable national resources can make disputes between countries more likely. In such contested areas, drones will enable governments to undertake strike missions or probe the responses of an adversary—actions they would be less inclined to take with manned platforms. According to the Central Intelligence Agency (CIA), there are approximately 430 bilateral maritime boundaries, most of which are not defined by formal agreements between the affected states.[26]

Beyond the cases of East Asia, other cross-border flashpoints for conflict where the low-risk proposition of drone strikes would be tempting include Russia in Georgia or Ukraine, Turkey in Syria, Sudan within its borders, and China on its western periphery. In 2013, a Chinese counternarcotics official revealed that his bureau had considered attempting to kill a drug kingpin named Naw Kham, who was hiding in a remote region in northeastern Myanmar, by using a drone carrying twenty kilograms of dynamite. "The plan was rejected, because the order was to catch him alive," the official recalled.[27] With armed drones, China might make the same calculation that the United States has made—that killing is more straightforward than capturing—in choosing to target ostensibly high-threat individuals with drone strikes. China's demonstrated willingness to employ armed drones against terrorists or criminals outside its borders could directly threaten U.S. allies in the region, particularly if the criterion China uses to define a terrorist does not align with that of the United States or its allies.

Domestically, governments may use armed drones to target their perceived internal enemies. Most emerging drone powers have experienced recent domestic unrest. Turkey, Russia, Pakistan, and China all have separatist or significant opposition movements (e.g., Kurds, Chechens, the Taliban, Tibetans, and Uighurs) that presented political and military challenges to their rule in recent history. These states already designate individuals from these groups as "terrorists," and reserve the right to use force against them. States possessing the lower risk—compared with other weapons platforms—capability of armed drones could use them more frequently in the service of domestic pacification, especially against time-sensitive targets that reside in mountainous, jungle, or other inhospitable terrain. Compared with typical methods used by military and police forces to counter insurgencies, criminals, or terrorists—such as ground troops and manned aircraft—unmanned drones provide significantly greater real-time intelligence through their persistent loiter time and responsiveness to striking an identified target.

INCREASED RISK OF MISPERCEPTION AND ESCALATION

Pushing limits in already unstable regions is complicated by questions raised regarding rules of engagement: how would states respond to an armed drone in what they contend is their sovereign airspace, and how would opposing sides respond to counter-drone tactics? Japanese defense officials claim that shooting down Chinese drones in what Japan contends is its airspace is more likely to occur than downing manned aircraft because drones are not as responsive to radio or pilot warnings, thereby raising the possibility of an escalatory response.[28] Alternatively, Japan might misidentify a Chinese manned fighter as an advanced drone and fire on it, especially if the aircraft's radar signature is not sufficiently distinctive or if combat drones routinely fly over the disputed area.

Thus, the additional risks associated with drone strikes, combined with the lack of clarity on how two countries would react to an attempted downing of a drone, create the potential for miscalculation and subsequent escalation. As U.S. Air Force commanders in South Korea noted, a North Korean drone equipped with chemical agents would not have to kill many or even any people on the peninsula to terrorize the population and escalate tensions.[29] This scenario points to the spiraling escalatory dynamic that could be repeated—likely intensified in the context of armed drones—in other tension-prone areas, such as the Middle East, South Asia, and Central and East Africa, where the mix of low-risk and ambiguous rules of engagement is a recipe for escalation. Not all of these contingencies directly affect U.S. interests, but they would affect treaty allies whose security the United States has an interest in maintaining. Compared with other weapons platforms, current practice repeatedly demonstrates that drones make militarized disputes more likely due to a decreased threshold for the use of force and an increased risk of miscalculation.

INCREASED RISK OF LETHALITY

The proliferation of armed drones will increase the likelihood of destabilizing or devastating one-off, high-consequence attacks. In March 2013, Senator Dianne Feinstein (D-CA) observed of drones: "In some respects it's a perfect assassination weapon. . . . Now we have a problem. There are all these nations that want to buy these armed drones. I'm strongly opposed to that."[30] The worst-case contingency for the use of armed drones, albeit an unlikely circumstance, would be to deliver weapons of mass destruction. Drones are, in many ways, the perfect vehicle for delivering biological and chemical agents.[31] A WMD attack, or even the assassination of a political leader, another troubling though unlikely circumstance, would have tremendous consequences for regional and international stability.

Deterring such drone-based attacks will depend on the ability of the United States and other governments to accurately detect and attribute them. Technical experts and intelligence analysts disagree about the extent to which this will be possible, but the difficulties lie in the challenges of detecting drones (they emit small radar, thermal, and electron signatures, and can fly low), determining who controlled it (they can be programmed to fly to a preset GPS coordinate), or assigning ownership to a downed system (they can be composed of commercial, off-the-shelf components).[32]

It is equally noteworthy that civilian officials or military commanders have almost always used armed drones in ways beyond their initially intended applications. Drones do not simply fulfill existing mission requirements; they create new and unforeseen ones, and will continue to do so in the future. Furthermore, U.S. officials would be misguided to view future uses of armed drones solely through the prism of how the United States has used them—for discrete military operations in relatively benign air-defense environments. The potential for misperception is compounded by the fact that few governments seeking or acquiring armed drones have publicly articulated any strategy for how they will likely use them. Conversely, the uncertainty about how other countries will use drones provides the United States with an opportunity to shape drone doctrines, especially for U.S. allies interested in procuring drones from U.S. manufacturers.

Proliferation Constraints and Incentives

Commercial drone applications advertised by companies such as Amazon give the illusion of a technology that is ubiquitous and inevitable. If drone technology indeed diffuses easily, then efforts to control its spread—whether through tight export controls or pressure on major producers to restrict drone transfers—are unnecessary and even misguided. In a world of rapid technology diffusion, countries will inevitably acquire the technology, and the United States has financial incentives to generate the greatest market share for its domestic industry. However, there are high barriers to entry for some countries that wish to join the armed drone market, which explain a low procurement rate despite the intrinsic advantages of drones and countries' stated ambitions to obtain them.

One factor is technological. Though rudimentary drones have existed for decades, there is a qualitative difference between earlier civilian and military surveillance models, and those used by the United States for strike missions. Drone strikes conducted by the United States require actionable intelligence (from human, signal, and imagery sources), sophisticated beyond line-of-sight communications, access to satellite bandwidth, and systems engineering—from internal fire control to ground control stations—that are presently beyond the reach of most states. Several countries with relatively advanced aerospace programs, including Russia, France, and Italy, have not been able to develop and deploy these capabilities. Though it is unlikely that countries will indigenously develop capabilities equal to that of the United States in the near term, most have the financial resources to purchase from the United States or Israel—including Pakistan, Turkey, Saudi Arabia, and the United Arab Emirates—and a desire to do so. Additionally, many are able to purchase or manufacture tactical short-range armed drones with limited firepower, which lack the additional system components required to carry out U.S. drone strikes, but can still have destabilizing effects.

A second factor is diplomatic. Drone strikes in foreign countries that allow for target intelligence collection necessitate a safe air environment and overflight rights, and require bilateral relationships to obtain host-nation basing rights for noncontiguous countries. U.S. drone strikes in Yemen and Somalia, for example, require airfields in Djibouti, Saudi Arabia, the Seychelles, and Ethiopia, secured with covert and overt aid and security commitments. (The United States does not conduct drone strikes from U.S. Navy ships, though it should be able to within five years.[33]) Few other countries will have reliable access to foreign airbases in coming years from which they can conduct lethal operations, and no other country will develop a blue-water navy capable of supporting intercontinental drone strikes for decades to come. Therefore, it is likely that most drone operations conducted by other countries within the coming years will be across borders or internal.

The third factor is domestic politics, which can constrain armed drone programs even in countries that have the ability to develop the technology. Whereas the U.S. targeted killing program has faced few domestic constraints, the politics of drones looks considerably different in some other countries. German politicians and military officials advocating for drones claim that there is a pressing military need. However, they have encountered intense opposition from a German public worried that the lethal capability would compromise the country's defense-only security norms, and increase the prospects for military interventions more generally. In late February, the European Parliament passed an unprecedented resolution, declaring, "Drone strikes outside a declared war by a state on the territory of another state without the consent of the latter or of the UN Security Council constitute a violation of international law and of the territorial integrity and sovereignty of that country."[34] The debates in Europe demonstrate how the prism through which opposing sides view armed drones is significantly influenced by their perception of the morality, legality, and necessity of U.S. drone strikes.

Though drones have lowered the threshold at which governments will authorize the use of force, they have not lowered it to zero. Risks remain, including diplomatic costs, for initiating even limited strikes with unmanned aerial systems. States that have armed drones but have not yet used them, such as Iran and China, have not been involved in militarized regional crises where such lethal capability would serve a strategic purpose. However, when they encounter heightened political or territorial tensions with neighboring countries, or face threats from across their borders, they will likely be more willing to use drones, or to counter them, in ways that are provocative and destabilizing—as demonstrated in the South and East China Seas. For example, Japan has stated it needs drones "to counter China's growing assertiveness at sea, especially when it comes to the Diaoyu/Senkaku Islands."[35] Traditional deterrence is applicable to the use of armed drones since most countries have an innate desire to avoid military escalation and conflict. However, because armed drones are used in destabilizing manners that other weapons platforms are not, there will be far more situations in which local military commanders must make decisions about employing force to counter drones, and the time for rational deterrence decision-making is significantly diminished.

Debates About Armed Drone Exports

A U.S. strategy to limit the proliferation of armed drones should consider the existing frameworks and principles that apply to their export. The 1987 Missile Technology Control Regime (MTCR) was intended to regulate nuclear-capable missiles and related technologies, including armed drones, and has strongly influenced U.S. policy debates about armed drone exports. Under the regime's guidelines, drones that can deliver a five hundred kilogram payload a minimum flight distance of three hundred kilometers are classified as Category I items, for which "there will be a strong presumption to deny such transfers." The United States was one of the seven original signatories to the MTCR, and maintains that its standards remain applicable to armed or unarmed Category I export considerations.

Given that the MTCR shapes and constrains U.S. drone exports, it is important to understand the challenges that it faces to remain relevant. First, it is a nonbinding international association, which the thirty-four signatory states interpret and implement at their discretion. Second, a number of armed drone–producing or aspiring countries are not members—Iran, Israel, China, India, and Pakistan—though Israel claims to be a unilateral adherent to its principles and China is a selective adherent.[36] Nonetheless, while the United States restricted its armed drone exports, Israel exported $4.6 billion in drone systems between 2005 and 2012, compared with less than $3 billion by the United States.[37] China has reportedly sold two of its smaller armed drones to the UAE and Pakistan, raising concerns about whether it would export its larger Predator-equivalent drone (the CH-4) to countries such as Iran.[38] Third, the Category I annex that addresses drones by payload and flight distance is arbitrary, since drones under these thresholds can conduct destabilizing probes into contested territories and lethal missions across borders, but would be exempt from this provision. For example, responding to requests from the UAE, General Atomics designed a

Category II version of the Predator (XP) for export. While this is preferable to exercising the rare exception to the presumption of denial and exporting a Category I system, it points to the arbitrary payload thresholds and previews ways in which the industry could—as technology improves and becomes lighter—export a Category I armed drone in the future.[39] It also suggests that an agreement organized around payload and range rather than mission type—lethal versus nonlethal—may decline in relevance as the technology evolves. Fourth, the "strong presumption of denial" clause suggests that member states should not export Category I items. The United States has refused to export such systems to Pakistan, the UAE, and Turkey, despite repeated requests from these countries. It has, however, exported such technologies to close allies, pointing to the potential for an erosion of any proliferation norm once other countries develop the capability to produce Category I items indigenously and begin exporting them.

Despite the MTCR's shortcomings, some U.S. officials and staffers worry that any attempt to rewrite the Category I restrictions would weaken nonproliferation norms more generally. This concern is based on the assumption that the global framework of multilateral treaties and norms on WMDs and their delivery vehicles are interconnected and reinforcing, and that weakening one weakens the regime as a whole. Moreover, in 2005, the Bush administration attempted to update and strengthen the language relating to drones, but could not get a consensus agreement, which is required for altering the MTCR. The Obama administration has undertaken a lengthy interagency review of U.S. drone export policies, and plans to publicly release some version of its findings in 2014.[40] Updated policies will need to balance the strong presumption of denial clause that garners broad, though incomplete, multilateral support, with more permissive armed drone exports to close allies and partners.

There are strong, unresolved disagreements among senior U.S. government officials over the future of U.S. exports and the policy tradeoffs. Officials who support leasing or selling most armed drone models contend that the weapons would play an essential role in building partnership capacity among allies and partners. Providing these countries with a wide range of lethal capabilities enables them to fight common enemies and allows the United States to promote their responsible use with monitoring and end-user verification agreements written into foreign military sales.[41] Officials further suggest that these cases would remain exceptions, and U.S. behavior may not be used as precedent for other countries, in which case the United States should not be hamstrung. Other government officials and staff argue that lowering the threshold for when the United States sells Category I drones would erode the norm created by other MTCR member states, as well as nonmembers, against such exports. In addition, some government officials worry that the introduction of new military capabilities would increase the likelihood of military escalation in unstable regions where there are unresolved territorial or boundary disputes, and believe that drones will be used irresponsibly—to intentionally probe adversaries, target civilians, or launch WMD attacks—or lead to arms racing.

Debates About Armed Drone Uses

A U.S. strategy that promotes the responsible use of armed drones should take into account the most contentious and unresolved issues surrounding them. In October 2013, two United Nations special rapporteurs published complementary reports condemning certain aspects of armed drones. Despite the generally critical tenor of the report, Christof Heyns, the UN special rapporteur on extrajudicial, summary, or arbitrary executions, conceded that "drones are here to stay" and are not, by their nature, illegal: "It is difficult to suggest that a weapon system is unlawful because a pilot is not on board."[42] Several countries nonetheless took the opportunity provided by the UN reports to critique the use of armed drones, with China pointing to the "blank space in international law" that is "subject to abuse."[43] These diplomatic challenges mirror the widespread foreign opposition to U.S. drone strikes, both in countries where they do not occur, and in Pakistan, for example, where a large majority of the population is opposed.[44] Public hatred among Pakistanis for U.S. drones strikes has put pressure on the government to oppose the United States on a range of non-drone issues, such as allowing the transportation of military equipment out of Afghanistan, in order to appease its domestic audience.[45]

The Obama administration has responded by arguing that all of its lethal counterterrorism strikes comply with domestic and international laws. The U.S. government's public position, as explained in speeches by senior officials, is that its use of lethal force is carried out in the context of an armed conflict with al-Qaeda and associated forces, to which the laws of armed conflict (international humanitarian law) apply. The United States also states that strikes are carried out pursuant to legitimate self-defense. U.S. officials have not clarified whether, or how, the United States also applies human rights law. Furthermore, after a long and deliberative interagency process, the White House published a summary of a presidential policy directive in May 2013 that describes the policies that the United States applies in its use of force operations outside areas of active hostilities. Regardless, many external observers have stated that legal boundaries remain unexplained and unclear. Even the former legal adviser to the State Department, Harold Koh, acknowledges there is still a "need for transparent, agreed-upon domestic and international legal process and standards."[46] In addition to the impacts of unclear U.S. policies and legal interpretations on accountability and democratic legitimacy, the lack of clarity also influences the capacity of the United States to set positive precedents for other states and to influence use-of-force norms.

To the extent that U.S. policy sets precedents for subsequent drone use, the lack of clarity about U.S. targeted killing policies should be addressed. For example, the Obama administration will not identify which terrorist groups can be lawfully targeted—only that targeted individuals are members of al-Qaeda or "associated forces"—because doing so would enhance the credibility of named groups, according to a Pentagon spokesperson.[47] Identifying these groups would increase transparency, reassuring other countries that the United States can justify who it targets. Additionally, this would give the United States leverage to call on other countries to explicitly define who they are targeting, rather than settle for vague descriptions, such as "associated forces."

Moreover, the Obama administration will not classify what belligerents' behavior or personal attributes make them a direct participant in hostilities, the latter of which covers the reported (though unacknowledged) controversial practice of signature strikes, in which individuals are targeted based on patterns of behavior—for example, presumed guilt for associating with a suspected terrorist. Similarly, the Obama administration will not publicize what procedures either the CIA or Department of Defense takes to prevent or mitigate harm to civilians, or to investigate incidents of civilian harm that occur during lethal operations. Such unresolved questions about what the United States considers to be the scope of the conflict and what procedures it has adopted to protect civilians have been raised repeatedly by allies to U.S. diplomats. Ambiguity regarding U.S. policies presents an opportunity for other countries to use drones with limited transparency and no clarification regarding processes and targeting. This lack of clarity also diminishes the United States' legitimacy by appearing hypocritical. Increasing transparency will enable the United States to take a leading role in shaping international norms on the use of drones by establishing credibility.

Though not all states will comply with norms, historical examples of weapons proliferation and use norms have been largely adhered to and in only rare instances not followed. For example, forty-five countries are capable of building nuclear weapons, but only nine have acquired them, which is fewer than the fifteen to twenty President John F. Kennedy predicted would exist in 1975. Similarly, although the United States refuses to sign the 1997 Mine Ban Treaty, it uses landmines only on the Korean Peninsula, where they are designed to be able to self-destruct in as little as four hours or at most fifteen days. Other classes of weapons have been eliminated, including biological weapons, or are in the hands of few countries, such as chemical weapons. Moreover, except in extremely rare circumstances, governments simply do not use force indiscriminately, nor claim they have the inherent right to do so.

Conclusions and Recommendations

The challenges highlighted by the U.S. experience using armed drones and the unresolved international debates suggest that the consequences of drone proliferation for other countries and its implications for U.S. interests will be considerable. As with other technology-driven global governance challenges, the longer the United States delays, the less influence it will have in shaping rules of the road, and the less likely it is that the largest possible coalition of emerging drone powers will agree to an arrangement addressing armed drones. Governing the use of armed drones will not require new treaties or international laws, because adequate legal instruments already exist, such as the MTCR, though it requires clarification. Additionally, the domestic politics of treaty ratification and the international politics of removing drones from the MTCR and creating a new treaty would only create more barriers to an international agreement on the governance of armed drones. Among countries with or pursuing armed drones, a more broadly accepted understanding of how specific legal terms are interpreted and applied is necessary, as well as more faithful and transparent adherence to them. In its remaining years, the Obama administration has the opportunity to play a significant role in what these rules of the road look like if it commits to help develop and shape them.

Some U.S. officials and analysts contend that the widespread proliferation of armed drones is inevitable, and that any efforts to influence their use will fail. This assertion disregards the diplomatic, domestic, political, and, for some, technological restraints that have limited the spread of other military capabilities, and the logistical, normative, and legal principles that affect whether and how they are used.

There are two broad policy decisions facing the Obama administration regarding the use and proliferation of armed drones. The most near-term policy decision is determining under what conditions the United States will export Category I unmanned systems. As the lead user of drones, the United States has the unique opportunity to determine which countries acquire these systems and to hold them accountable for how they use those drones. Toward this end, the following principles should guide all related Category I exports:

- a commitment to peacefully resolving all outstanding border or maritime disputes;
- a commitment to brokering domestic political disputes peacefully;
- a demonstrated history of protecting civilians from harm caused by uses of other weapons platforms;
- a commitment to human rights protections;
- a commitment to transparency in how armed drones are employed—including support of strategies, doctrine, and the applicable legal framework;
- a commitment to WMD and WMD-delivery nonproliferation; and
- a demonstrated history of rigorous adherence to end-user agreements for U.S.-supplied weapons, including unauthorized third-party transfers.

The United States is currently confronted with a concrete example of how these principles should be applied: the Obama administration is facing a near-term decision about exporting Category I armed Reaper drones to Italy. Given Italy's general support for such principles, the Obama administration should waive the strong presumption of denial to provide such weapons platforms to Italy, which would enhance the military capabilities of a close ally, promote interoperability between American and Italian military forces, and provide industrial support for a U.S. firm that has seen a decrease in domestic sales since the Pentagon no longer requires, nor can afford, to operate the large number of drones deployed during the surge in Afghanistan. These steps should be implemented in stages to ensure that the drones will be used responsibly, by first leasing a small number with a monitoring and evaluation program built in, and then selling outright a few years later.

Given that it is unrealistic for the United States to get consensus support for modifying the MTCR, any exports of Category I unmanned systems should be concomitant with public confidence-building measures concerning each individual sale. Specifically, the United States should provide detailed descriptions of armed drone exports in its annual report to the UN Register of Conventional Arms, which is not currently required for member states. This would provide transparency of U.S. weapons development and decrease uncertainty among potential adversaries about its capabilities, subsequently decreasing the likelihood of an arms race.

The second broad policy decision facing the Obama administration is the establishment of norms and practices for how and in what situations armed drones should be used. Customary international law is based in part on state practice, which also influences other countries' interpretations of treaties. Given that the United States is the lead actor and exemplar of drone use, its interpretation of international law, public articulation of its position, and future behavior will set a precedent on which other countries are likely to base their own behavior. The actions of the United States would serve as a benchmark against which others are judged, and therefore provide legitimacy for and reduce the political and diplomatic costs of other countries emulating U.S. practices. This does not entail revealing information about the sources and methods for targeted strikes, which is a legitimate concern for many U.S. officials, particularly in the intelligence community. This sort of transparency has been successfully employed on other highly sensitive national security issues—in 2010, as part of the Nuclear Posture Review, the United States outlined specifics for the circumstances under which it would employ nuclear weapons, what force posture it would maintain to deliver them, and even how many nuclear warheads were in its arsenal.

Though the United States has made initial efforts to increase transparency, especially in the context of providing information to its own citizens, it should extend that transparency to all its targeting decisions. In February 2014, Director of National Intelligence James Clapper said, regarding armed drones, "I would hope, as other countries acquire similar capabilities, that they follow the model that we have for the care and precision that we exercise." Though this supposed model could serve as a good precedent, it remains largely a secret, thereby leaving a precedent of minimal transparency and a lack of justification for drone strikes. A guiding principle for how the United States describes and clarifies its drone operations should be based on the type and specificity of information it wants to see used by other armed drone states. The Obama administration has provided transparency for some issues, such as the November 2012 Department of Defense directive mandating "appropriate levels of human judgment over the use of force" by autonomous and semiautonomous weapons systems.[48] In other areas it remains opaque, particularly in regard to the supposed "strong preference" for capturing suspected terrorists whenever possible rather than killing them, given that the available information overwhelmingly concludes the opposite is taking place.

The United States should be more specific in terms of how it applies international humanitarian law and self-defense law to its policy of armed drone strikes—and whether international human rights law applies. Legal constraints signal the United States' preference for how it wants other countries to conduct drone strikes in the future and, although not all states will follow U.S. precedent, give the United States leverage to shape norms and condemn the actions of other states. Although established norms and practices would limit the use of armed drones, self-defense and regular military missions would still be protected under international law. Additionally, the United States should explain how it interprets "continuing and imminent" threats in the context of the relevant body of international law, a primary point of contention and confusion among European allies. The United States takes a more expansive view of self-defense and imminence than its allies, not just with respect to drones and targeting individuals, but also to invading countries, and there may be consequences to this broader discretion regarding international law. Unless the United States accepts some constraints and leads the way in developing them, it will face greater operational constraints—stemming from decreased counterterrorism cooperation, denial of overflight or basing rights, or diplomatic pressure—as a result of future backlash.

Beyond these two broad sets of recommendations, there are several subsequent policy recommendations for the Obama administration:

- Task the intelligence community to publish an unclassified survey of the current and future trends of unmanned military technologies—including ground, sea, and autonomous systems—as it does for ballistic missiles and WMDs. This would clarify the state of proliferation and use for the U.S. government and domestic drone manufacturers, and inform changes to U.S. strategic guidance, military posture, and export principles.
- Commission an unclassified study by a federally funded research institution to assess how unmanned aerial systems have been employed in destabilizing settings and identify the most likely potential future missions of drones that run counter to U.S. interests.
- Direct administration officials to testify—for the first time—before Senate and House Foreign Relations Committees hearings on the

unresolved questions over what principles and criteria should guide armed and unarmed drone exports.

- Appoint a high-level panel of outside experts to review U.S. government policies on targeting decisions and their transparency and potential effect on emerging proliferators, and propose reforms based on the President's Review Group on Intelligence and Communication Technologies.
- Encourage U.S. drone manufacturers, who seek lower barriers to export, to form an association—comparable to the World Association of Nuclear Operators—to promote industry growth through transparency and risk management.
- Redouble Track I and Track II efforts to address growing apprehension among European governments and nongovernmental organizations about fully autonomous lethal drones, which the U.S. military has declared it is not pursuing.
- Reevaluate the MTCR's Category I versus Category II distinctions that are currently based on arbitrary distinctions of payloads over five hundred kilograms and ranges not exceeding three hundred kilometers, which excludes potentially lethal and destabilizing drones. Consider categories based on mission type (e.g., lethal versus nonlethal).
- Review whether the strong presumption of denial clause in the MTCR offers clear guidelines for states, and investigate cases where member states have overcome that presumption and exported Category I items to measure the consequences for drone proliferation norms.
- Formally request that non-signatories to the MTCR—including China, Iran, Pakistan, and India—issue public declarations of unilateral support for the principles and regulations contained in the regime.
- Begin formal discussions with Chinese government leaders to increase transparency and predictability for both American and Chinese drone exports, as part of the U.S.-China Strategic and Economic Dialogue process.
- Form an international working group composed of emerging proliferation powers to identify the consequences of future drone technologies, dual-use concerns (i.e., nonlethal drones that can be made lethal), and the viability of so-called trigger lists along the lines of the Zangger Committee lists for nuclear technology.

Endnotes

1. "Israeli Unmanned Aerial Systems," Frost & Sullivan, May 20, 2013.
2. Andrew Robathan, "Written Answers to Questions," UK Parliament, September 5, 2013.
3. Barbara Opall-Rome, "Israeli Missiles Down Armed Hizbollah UAVs," *Defense News*, October 2, 2006.
4. "Precisely Wrong: Gaza Civilians Killed by Israeli Drone-Launched Missiles," Human Rights Watch, 2009; Ashraf Sweilman, "Officials: Israeli Drone Strike Kills Five in Egypt," *Washington Post*, August 9, 2013.
5. "Iran Unveils 'Indigenous' Drone With 2,000km Range," BBC News, September 26, 2012.
6. Robert Karnoil, "Unmanned Aircraft Poised to Fill Asian Skies," *Straits Times*, May 28, 2012.
7. *Unmanned Aerial Vehicle (UAV) Market (2013 – 2018)*, MarketsandMarkets, June 2013.
8. "UAVs Capable of Launching Weapons Soon: DRDO Chief," *The Hindu*, August 27, 2013.
9. Sebastian Abbot, "Pakistan Struggles in Race to Develop Armed Drones," Associated Press, November 17, 2012.
10. Keir Simmons and Gil Aegerter, "The Race is On: Manufacturer Sets its Sights on Market for Armed Drones," NBC News, May 28, 2013.
11. Rob O'Gorman and Chris Abbott, "Remote Control War: Unmanned Combat Air Vehicles in China, India, Iran, Israel, Russia, and Turkey," Open Briefing, September 20, 2013.
12. Duncan Geere, "Europe's Stealth Drone, Neuron, Takes Flight," Wired, December 2, 2012.
13. Amy Svitak, "Falling Short," *Aviation Week and Space Technology*, December 30, 2013, pp. 19–20.
14. Mark Dodd, "Armed Drones a Future Option for our Forces," *Australian*, May 29, 2012.
15. Tom Lawson, remarks at the Center for Strategic and International Studies, November 26, 2013.
16. Norton A. Schwartz, "Pragmatic Steps for Global Security," remarks at the Stimson Center, May 1, 2012.
17. Interviews with U.S. government officials.
18. Certain missiles fired from other weapons platforms can also be diverted, but as a matter of practice, they rarely are. Missiles fired from drones are diverted much more frequently.
19. Daniel Byman, "Why Drones Work: The Case for Washington's Weapon of Choice," *Foreign Affairs*, July/August 2013.

20. Lotta Themner and Peter Wallensteen, "Armed Conflicts, 1946–2012," *Journal of Peace Research*, 50:4, July 2, 2013, p. 510.
21. Bill Roggio and Alexander Mayer, "Charting the Data for U.S. Airstrikes in Pakistan, 2004–2014," *Long War Journal*, last updated December 25, 2013; Bill Roggio and Bob Barry, "Charting the Data for U.S. Airstrikes in Yemen, 2002–2014," *Long War Journal*, last updated June 5, 2014; "Yemen: Reported U.S. Covert Action 2001–2011," Bureau of Investigative Journalism, March 29, 2012; "Yemen: Reported U.S. Covert Action 2012," Bureau of Investigative Journalism, May 8, 2012; "Yemen: Reported U.S. Covert Actions 2013," Bureau of Investigative Journalism, January 3, 2013; "Yemen: Reported U.S. Covert Actions 2014," Bureau of Investigative Journalism, updated June 4, 2014; Mark Mazzetti, *The Way of the Knife: The CIA, a Secret Army, and a War at the Ends of the Earth* (New York: Penguin Press, 2013), p. 134.
22. Average based on ranges provided by New America Foundation (NAF) and Bureau of Investigative Journalism through June 2014.
23. Greg Jaffe, "Former Defense Secretary Gates Warns Against Lure of Drone Warfare," *Washington Post*, October 23, 2013.
24. Jason Koebler, "Obama: Administration Saw Drone Strikes as 'Cure-All' for Terrorism," *U.S. News and World Report*, May 23, 2013.
25. John O. Brennan, "The Ethics and Efficacy of the President's Counterterrorism Strategy," Wilson Center, April 30, 2012; and Mark Bowden, *The Finish: The Killing of Osama Bin Laden* (New York: Atlantic Monthly Press, 2012), p. 262.
26. *The World Factbook 2013–14*, Central Intelligence Agency, 2013.
27. Liu Chang, "Manhunt for Deadly Drug Kingpin," *Global Times*, February 19, 2013.
28. Itsunori Onodera, "Press Conference by the Defense Minister," Japan Ministry of Defense, September 10, 2013; and "Defense Ministry Working on Protocol to Shoot Down Encroaching Drones," *Asahi Shimbun*, October 2, 2013.
29. Interview with senior U.S. Air Force official stationed in South Korea, April 2014.
30. Breanna Edwards, "Dianne Feinstein: Time to Set Drone Rules," *Politico*, March 7, 2013. Indeed Feinstein has been a strong opponent of selling armed variants of drones, including to NATO allies.
31. Dennis M. Gormley, "Limiting the Unintended Consequences of Unmanned Air System Proliferation," *Whitehead Journal of Diplomacy & International Relations*, Winter/Spring 2013, pp. 67–79.
32. David Sanger and William Broad, "Pentagon Study Finds Agencies Ill Equipped to Detect Foreign Nuclear Efforts," *New York Times*, January 24, 2014.
33. Christopher Cavas, "X-47B Successfully Lands on Carrier," *Defense News*, July 15, 2013.
34. "European Parliament resolution on the use of armed drones (2014/2567(RSP))," European Parliament, February 25, 2014.
35. "Report: Japan Mulls U.S. Spy Drones for Marine Surveillance," Agence France-Presse, January 4, 2013.
36. Dennis Gormley, *Missile Contagion: Cruise Missile Proliferation and the Threat to International Security* (Annapolis, MD: Naval Institute Press, 2010), pp. 61–62 and 138–39.
37. Michael Blades, "Israeli Unmanned Aerial Systems," Frost & Sullivan, March 14, 2013. Data on armed drones is limited, but market-wide data is indicative of likely trends.
38. Bill Gertz, "Red Dawn: Communist China Stepping up Drone Deployment," *Washington Times*, March 26, 2013.
39. Chris Pocock, "Predator UAV Cleared for Wider Export," *AIN Defense Perspective*, March 4, 2011.
40. Interviews with current and former U.S. officials. On January 15, 2013, the White House released Presidential Policy Directive/PPD-27, "United States Conventional

Arms Transfer Policy," which provides generic guidance and does not mention unmanned aerial systems or norm setting.

41. In July 2012, the Government Accountability Office suggested that the export policies then had important gaps, in terms of end-user verification of how the UAVs were used and how this information was coordinated across agencies. Some of the affected the government agencies claim that these shortcomings have been adequately addressed.
42. Christof Heyns, "Statement by Special Rapporteur on Extrajudicial, Summary or Arbitrary Executions," UN General Assembly, October 25, 2013.
43. Ed Pilkington and Ryan Devereaux, "U.S. Defends Drone Strikes as 'Necessary and Just' in Face of UN Criticism," *Guardian*, October 25, 2013.
44. "Global Opinion of Obama Slips, International Policies Faulted," Pew Research Center, June 13, 2012; "America's Image Remains Poor: Concern About Extremist Threat Slips in Pakistan," Pew Research Center, July 29, 2010.
45. Arsla Jawaid, "Drones May Be Sharif's Only Weapon in Talks With Obama," *Guardian*, October 23, 2013.
46. Harold Koh, remarks at Oxford Union, "How to End the Forever War," May 7, 2013.
47. Cora Currier, "Who Are We at War With? That's Classified," *ProPublica*, July 26, 2013.
48. Department of Defense, "Autonomy in Weapon Systems," directive 3000.09, November 21, 2012.

About the Authors

Micah Zenko is the Douglas Dillon fellow in the Center for Preventive Action at the Council on Foreign Relations. Previously, he worked for five years at Harvard University's Kennedy School of Government, and in Washington, DC, at the Brookings Institution, Congressional Research Service, and State Department's Office of Policy Planning. Zenko has published on a range of national security issues, including articles in *Foreign Affairs*, *Journal of Strategic Studies*, *Defense and Security Analysis*, and *Annals of the American Academy of Political and Social Science*, as well as op-eds in the *Washington Post*, *Los Angeles Times*, *Chicago Tribune*, and *New York Times*. He writes the CFR blog *Politics, Power, and Preventive Action*, which covers U.S. national security policy, international security, and conflict prevention. He is the author of three Council Special Reports: *Partners in Preventive Action: The United States and International Institutions*; *Toward Deeper Reductions in U.S. and Russian Nuclear Weapons*; and *Enhancing U.S. Preventive Action*. He also wrote *Between Threats and War: U.S. Discrete Military Operations in the Post–Cold War World*. Zenko received a PhD in political science from Brandeis University.

Sarah Kreps is a Stanton nuclear security fellow at the Council on Foreign Relations. She is on leave from Cornell University, where she is an assistant professor in the department of government and an adjunct professor at Cornell Law School. She is the author of *Coalitions of Convenience: United States Military Interventions after the Cold War* and the forthcoming book *Drone Warfare*. She has published in international security journals, including the *American Political Science Review*, *Journal of Politics*, *Security Studies*, and *Political Science Quarterly*, as well as op-eds in the *New York Times*, *Washington Post*, *USA Today*, and *Herald Tribune*. Kreps received a BA from Harvard University, MSc from Oxford University, and PhD from Georgetown University. She was previously an acquisitions and foreign area officer in the U.S. Air Force.

Advisory Committee for *Limiting Armed Drone Proliferation*

Brian C. Babcock-Lumish
U.S. Military Academy

Daniel Byman
Edmund A. Walsh School of Foreign Service

Robert M. Chesney
University of Texas School of Law

Stephen J. Del Rosso Jr.
Carnegie Corporation of New York

Ian Easton
Project 2049 Institute

James T. Golby
U.S. Military Academy

Dennis M. Gormley
University of Pittsburgh

Michael C. Horowitz
University of Pennsylvania

Robert O. Keohane
Princeton University

Sarah Knuckey
NYU School of Law

Matthew Leatherman
Citizen Soldier Support Program

Joanne Mariner
Amnesty International

Daniel Markey, *ex officio*
Council on Foreign Relations

David A. Merkel
Summit International Advisors

Scott Osterling
Cameron International

Ely Ratner
Center for a New American Security

Andrew Shapiro
Beacon Global Strategies

Paul Stares, *ex officio*
Council on Foreign Relations

Ward Thomas
College of the Holy Cross

Erik Voeten
Edmund A. Walsh School of Foreign Service

Graham Warwick
Aviation Week and Space Technology

Steven J. Zaloga
The Teal Group

This report reflects the judgments and recommendations of the authors. It does not necessarily represent the views of members of the advisory committee, whose involvement in no way should be interpreted as an endorsement of the report by either themselves or the organizations with which they are affiliated.

Center for Preventive Action Advisory Committee

Mission Statement of the Center for Preventive Action

The Center for Preventive Action (CPA) seeks to help prevent, defuse, or resolve deadly conflicts around the world and to expand the body of knowledge on conflict prevention. It does so by creating a forum in which representatives of governments, international organizations, nongovernmental organizations, corporations, and civil society can gather to develop operational and timely strategies for promoting peace in specific conflict situations. The center focuses on conflicts in countries or regions that affect U.S. interests, but may be otherwise overlooked; where prevention appears possible; and when the resources of the Council on Foreign Relations can make a difference. The center does this by

- Issuing Council Special Reports to evaluate and respond rapidly to developing conflict situations and formulate timely, concrete policy recommendations that the U.S. government and international and local actors can use to limit the potential for deadly violence.
- Engaging the U.S. government and news media in conflict prevention efforts. CPA staff members meet with administration officials and members of Congress to brief on CPA findings and recommendations; facilitate contacts between U.S. officials and important local and external actors; and raise awareness among journalists of potential flashpoints around the globe.
- Building networks with international organizations and institutions to complement and leverage the Council's established influence in the U.S. policy arena and increase the impact of CPA recommendations.
- Providing a source of expertise on conflict prevention to include research, case studies, and lessons learned from past conflicts that policymakers and private citizens can use to prevent or mitigate future deadly conflicts.

Council Special Reports

Published by the Council on Foreign Relations

Reorienting U.S. Pakistan Strategy: From Af-Pak to Asia
Daniel S. Markey; CSR No. 68, January 2014

Afghanistan After the Drawdown
Seth G. Jones and Keith Crane; CSR No. 67, November 2013
A Center for Preventive Action Report

The Future of U.S. Special Operations Forces
Linda Robinson; CSR No. 66, April 2013

Reforming U.S. Drone Strike Policies
Micah Zenko; CSR No. 65, January 2013
A Center for Preventive Action Report

Countering Criminal Violence in Central America
Michael Shifter; CSR No. 64, April 2012
A Center for Preventive Action Report

Saudi Arabia in the New Middle East
F. Gregory Gause III; CSR No. 63, December 2011
A Center for Preventive Action Report

Partners in Preventive Action: The United States and International Institutions
Paul B. Stares and Micah Zenko; CSR No. 62, September 2011
A Center for Preventive Action Report

Justice Beyond The Hague: Supporting the Prosecution of International Crimes in National Courts
David A. Kaye; CSR No. 61, June 2011

The Drug War in Mexico: Confronting a Shared Threat
David A. Shirk; CSR No. 60, March 2011
A Center for Preventive Action Report

UN Security Council Enlargement and U.S. Interests
Kara C. McDonald and Stewart M. Patrick; CSR No. 59, December 2010
An International Institutions and Global Governance Program Report

Congress and National Security
Kay King; CSR No. 58, November 2010

Toward Deeper Reductions in U.S. and Russian Nuclear Weapons
Micah Zenko; CSR No. 57, November 2010
A Center for Preventive Action Report

Internet Governance in an Age of Cyber Insecurity
Robert K. Knake; CSR No. 56, September 2010
An International Institutions and Global Governance Program Report

From Rome to Kampala: The U.S. Approach to the 2010 International Criminal Court Review Conference
Vijay Padmanabhan; CSR No. 55, April 2010

Strengthening the Nuclear Nonproliferation Regime
Paul Lettow; CSR No. 54, April 2010
An International Institutions and Global Governance Program Report

The Russian Economic Crisis
Jeffrey Mankoff; CSR No. 53, April 2010

Somalia: A New Approach
Bronwyn E. Bruton; CSR No. 52, March 2010
A Center for Preventive Action Report

The Future of NATO
James M. Goldgeier; CSR No. 51, February 2010
An International Institutions and Global Governance Program Report

The United States in the New Asia
Evan A. Feigenbaum and Robert A. Manning; CSR No. 50, November 2009
An International Institutions and Global Governance Program Report

Intervention to Stop Genocide and Mass Atrocities: International Norms and U.S. Policy
Matthew C. Waxman; CSR No. 49, October 2009
An International Institutions and Global Governance Program Report

Enhancing U.S. Preventive Action
Paul B. Stares and Micah Zenko; CSR No. 48, October 2009
A Center for Preventive Action Report

The Canadian Oil Sands: Energy Security vs. Climate Change
Michael A. Levi; CSR No. 47, May 2009
A Maurice R. Greenberg Center for Geoeconomic Studies Report

The National Interest and the Law of the Sea
Scott G. Borgerson; CSR No. 46, May 2009

Lessons of the Financial Crisis
Benn Steil; CSR No. 45, March 2009
A Maurice R. Greenberg Center for Geoeconomic Studies Report

Global Imbalances and the Financial Crisis
Steven Dunaway; CSR No. 44, March 2009
A Maurice R. Greenberg Center for Geoeconomic Studies Report

Eurasian Energy Security
Jeffrey Mankoff; CSR No. 43, February 2009

Preparing for Sudden Change in North Korea
Paul B. Stares and Joel S. Wit; CSR No. 42, January 2009
A Center for Preventive Action Report

Averting Crisis in Ukraine
Steven Pifer; CSR No. 41, January 2009
A Center for Preventive Action Report

Congo: Securing Peace, Sustaining Progress
Anthony W. Gambino; CSR No. 40, October 2008
A Center for Preventive Action Report

Deterring State Sponsorship of Nuclear Terrorism
Michael A. Levi; CSR No. 39, September 2008

China, Space Weapons, and U.S. Security
Bruce W. MacDonald; CSR No. 38, September 2008

Sovereign Wealth and Sovereign Power: The Strategic Consequences of American Indebtedness
Brad W. Setser; CSR No. 37, September 2008
A Maurice R. Greenberg Center for Geoeconomic Studies Report

Securing Pakistan's Tribal Belt
Daniel S. Markey; CSR No. 36, July 2008 (web-only release) and August 2008
A Center for Preventive Action Report

Avoiding Transfers to Torture
Ashley S. Deeks; CSR No. 35, June 2008

Global FDI Policy: Correcting a Protectionist Drift
David M. Marchick and Matthew J. Slaughter; CSR No. 34, June 2008
A Maurice R. Greenberg Center for Geoeconomic Studies Report

Dealing with Damascus: Seeking a Greater Return on U.S.-Syria Relations
Mona Yacoubian and Scott Lasensky; CSR No. 33, June 2008
A Center for Preventive Action Report

Climate Change and National Security: An Agenda for Action
Joshua W. Busby; CSR No. 32, November 2007
A Maurice R. Greenberg Center for Geoeconomic Studies Report

Planning for Post-Mugabe Zimbabwe
Michelle D. Gavin; CSR No. 31, October 2007
A Center for Preventive Action Report

The Case for Wage Insurance
Robert J. LaLonde; CSR No. 30, September 2007
A Maurice R. Greenberg Center for Geoeconomic Studies Report

Reform of the International Monetary Fund
Peter B. Kenen; CSR No. 29, May 2007
A Maurice R. Greenberg Center for Geoeconomic Studies Report

Nuclear Energy: Balancing Benefits and Risks
Charles D. Ferguson; CSR No. 28, April 2007

Nigeria: Elections and Continuing Challenges
Robert I. Rotberg; CSR No. 27, April 2007
A Center for Preventive Action Report

The Economic Logic of Illegal Immigration
Gordon H. Hanson; CSR No. 26, April 2007
A Maurice R. Greenberg Center for Geoeconomic Studies Report

The United States and the WTO Dispute Settlement System
Robert Z. Lawrence; CSR No. 25, March 2007
A Maurice R. Greenberg Center for Geoeconomic Studies Report

Bolivia on the Brink
Eduardo A. Gamarra; CSR No. 24, February 2007
A Center for Preventive Action Report

After the Surge: The Case for U.S. Military Disengagement From Iraq
Steven N. Simon; CSR No. 23, February 2007

Darfur and Beyond: What Is Needed to Prevent Mass Atrocities
Lee Feinstein; CSR No. 22, January 2007

Avoiding Conflict in the Horn of Africa: U.S. Policy Toward Ethiopia and Eritrea
Terrence Lyons; CSR No. 21, December 2006
A Center for Preventive Action Report

Living with Hugo: U.S. Policy Toward Hugo Chávez's Venezuela
Richard Lapper; CSR No. 20, November 2006
A Center for Preventive Action Report

Reforming U.S. Patent Policy: Getting the Incentives Right
Keith E. Maskus; CSR No. 19, November 2006
A Maurice R. Greenberg Center for Geoeconomic Studies Report

Foreign Investment and National Security: Getting the Balance Right
Alan P. Larson and David M. Marchick; CSR No. 18, July 2006
A Maurice R. Greenberg Center for Geoeconomic Studies Report

Challenges for a Postelection Mexico: Issues for U.S. Policy
Pamela K. Starr; CSR No. 17, June 2006 (web-only release) and November 2006

U.S.-India Nuclear Cooperation: A Strategy for Moving Forward
Michael A. Levi and Charles D. Ferguson; CSR No. 16, June 2006

Generating Momentum for a New Era in U.S.-Turkey Relations
Steven A. Cook and Elizabeth Sherwood-Randall; CSR No. 15, June 2006

Peace in Papua: Widening a Window of Opportunity
Blair A. King; CSR No. 14, March 2006
A Center for Preventive Action Report

Neglected Defense: Mobilizing the Private Sector to Support Homeland Security
Stephen E. Flynn and Daniel B. Prieto; CSR No. 13, March 2006

Afghanistan's Uncertain Transition From Turmoil to Normalcy
Barnett R. Rubin; CSR No. 12, March 2006
A Center for Preventive Action Report

Preventing Catastrophic Nuclear Terrorism
Charles D. Ferguson; CSR No. 11, March 2006

Getting Serious About the Twin Deficits
Menzie D. Chinn; CSR No. 10, September 2005
A Maurice R. Greenberg Center for Geoeconomic Studies Report

Both Sides of the Aisle: A Call for Bipartisan Foreign Policy
Nancy E. Roman; CSR No. 9, September 2005

Forgotten Intervention? What the United States Needs to Do in the Western Balkans
Amelia Branczik and William L. Nash; CSR No. 8, June 2005
A Center for Preventive Action Report

A New Beginning: Strategies for a More Fruitful Dialogue with the Muslim World
Craig Charney and Nicole Yakatan; CSR No. 7, May 2005

Power-Sharing in Iraq
David L. Phillips; CSR No. 6, April 2005
A Center for Preventive Action Report

Giving Meaning to "Never Again": Seeking an Effective Response to the Crisis in Darfur and Beyond
Cheryl O. Igiri and Princeton N. Lyman; CSR No. 5, September 2004

Freedom, Prosperity, and Security: The G8 Partnership with Africa: Sea Island 2004 and Beyond
J. Brian Atwood, Robert S. Browne, and Princeton N. Lyman; CSR No. 4, May 2004

Addressing the HIV/AIDS Pandemic: A U.S. Global AIDS Strategy for the Long Term
Daniel M. Fox and Princeton N. Lyman; CSR No. 3, May 2004
Cosponsored with the Milbank Memorial Fund

Challenges for a Post-Election Philippines
Catharin E. Dalpino; CSR No. 2, May 2004
A Center for Preventive Action Report

Stability, Security, and Sovereignty in the Republic of Georgia
David L. Phillips; CSR No. 1, January 2004
A Center for Preventive Action Report

Note: Council Special Reports are available for download from CFR's website, www.cfr.org. For more information, email publications@cfr.org.

Made in the USA
Monee, IL
13 September 2026

62977346R00030